JOSE CARLOS POMA LOAYZA
CARLOS POMA RAMOS
GUSTAVO RICHARD VELIZ ESPINOZA

POLYACRYLAMIDE POLYMER (PAM) IN SOIL STABILIZATION

JOSE CARLOS POMA LOAYZA
CARLOS POMA RAMOS
GUSTAVO RICHARD VELIZ ESPINOZA

POLYACRYLAMIDE POLYMER (PAM) IN SOIL STABILIZATION

A WP STUDY FOR ROAD SOIL STABILIZATION

Imprint

Any brand names and product names mentioned in this book are subject to trademark, brand or patent protection and are trademarks or registered trademarks of their respective holders. The use of brand names, product names, common names, trade names, product descriptions etc. even without a particular marking in this work is in no way to be construed to mean that such names may be regarded as unrestricted in respect of trademark and brand protection legislation and could thus be used by anyone.

Cover image: www.ingimage.com

This book is a translation from the original published under ISBN 978-613-9-40669-2.

Publisher:
Sciencia Scripts
is a trademark of
Dodo Books Indian Ocean Ltd. and OmniScriptum S.R.L publishing group

120 High Road, East Finchley, London, N2 9ED, United Kingdom
Str. Armeneasca 28/1, office 1, Chisinau MD-2012, Republic of Moldova, Europe
Printed at: see last page
ISBN: 978-620-7-77740-2

INDEX

SUMMARY

The general problem *posed in this research was: How does the application of the polymer polyacrylamide (PAM) influence soil stabilisation on local roads in the Province of Angaraes?-* ***The general objective*** *of the study was: To determine the influence of the application of the polymer polyacrylamide (PAM) in the stabilisation of the soil of the local roads of the Province of Angaraes - Huancavelica, and the* ***general hypothesis****: The application of the polymer polyacrylamide (PAM) in the stabilisation of the soil of the local roads of the Province of Angaraes - Huancavelica will influence the physical-mechanical properties. The general method of investigation was scientific, of an applicative type, descriptive-explanatory research level, quasi-experimental design, the population was all the local roads of the Province of Angaraes and the non-probabilistic sample was constituted by the local road Section Lircay - Jatumpata - Mitoccasa -Dv. Ccollpa - Pitinpata - Perccapampa, which has a length of 16 + 670 km. The research concluded that by applying the polymer polyacrylamide (PAM) to the quarry material it influences its physical-mechanical properties, improving its useful life and reducing its intervention interval (periodic maintenance).*

Keywords: *Polymer, Polyacrylamide (PAM), Soil stabilisation, Service levels.*

INTRODUCTION

Low-volume, unpaved roads in Peru are mostly deteriorated compared to other countries due to the high cost of investment for their maintenance. In the Province of Angaraes, due to heavy rainfall in the months of December to March, the platforms of its local roads deteriorate rapidly, which is why the present study entitled **"APPLICATION OF POLYACRYLAMIDE POLYMER (PAM) IN THE STABILISATION OF THE SOIL IN THE ROADS OF THEPROVINCIA DE ANGARAES - HUANCAVELICA"** for the improvement of the physical-mechanical properties of the soil.

The present research is made up of five chapters, which are as follows:

CHAPTER I: The first chapter describes the problem of t h e research, where the general problem and the other specific problems are formulated, justification, delimits and subsequently determines the objectives.

CHAPTER II: In the second chapter the theoretical framework is developed, where the national and international background is illustrated with the definitions of the terms; at this point the hypothesis and the variables of the research, which are important for the development of this research, are set out.

CHAPTER III: The third chapter is part of the research methodology, which describes the research method, type of research, level of research, research design, population, sample, data collection techniques and instruments.

CHAPTER IV: The fourth chapter details the results of the investigation, which are obtained from the soil mechanics laboratory tests, as well as the results of the research. The main quarry as well as the combination with polyacrylamide polymer and the assessment of the level of service of the local road.

CHAPTER V: The fifth chapter contains the discussion of the results where the results obtained are analysed and evaluated. This is followed by the conclusions of the research, the recommendations, the bibliographical references and finally the annexes. Bach. JOSE CARLOS POMA LOAYZA

CHAPTER I

THE PROBLEM OF RESEARCH

1.1. Approach of the problem

In Latin American countries, there is a deficient transport infrastructure in terms of unpaved roads, which are deteriorated due to the little maintenance required to maintain service levels, compared to European road infrastructure.

In Peru, low traffic volume unpaved roads are of great importance for local, regional and national development, as they have 90232 km of road surface at the national level. One of the most important problems at the national level is the high cost of investment for the maintenance of these roads.

RED VIAL (N° Rutas)	EXISTENTE POR TIPO DE SUPERFICIE DE RODADURA				PROYECTADA	TOTAL	
	PAVIMENTADA	NO PAVIMENTADA		SUB TOTAL			
		Afirmada	Sin Afirmar				
NACIONAL (130)	14,747.74	7,631.51	2,214.16	24,593.40	1,901.29	26,494.69	17.7%
DEPARTAMENTAL (386)	2,339.72	14,263.37	7,632.04	24,235.12	4,794.49	29,029.62	19.4%
VECINAL (6,244)	1,611.10	19,231.34	71,001.39	91,843.83	2,291.83	94,135.66	62.9%
TOTAL	18,698.56	41,126.21	80,847.59	140,672.36	8,987.61	149,659.97	100%

Figure 1. National Road System in Peru Source: General Directorate of Roads and Railways.

The evaluation of a low traffic volume road surface is based on the cost-benefit analysis carried out by the state and determines the feasibility of the project, most of which are not sustainable, some of which have operation and maintenance, these operating costs are very high. The Province of Angaraes has 1500 km of local roads, only 10% of which are maintained by the Provincial Road Institute of Angaraes. A fundamental problem is the maintenance of their level of service, which has repercussions on the high costs of maintenance and investment by the state. These local roads in the Province of Angaraes have an average IMDA (Average Daily Annual Daily Index) of 20 to 60 vehicles/day, which prevents further investment in the maintenance or rehabilitation of these roads.In the District of Lircay in the months of December, January, February and March, due to heavy rainfall, the platforms of its local roads deteriorate rapidly, forming potholes, rutting, rutting and loss of pumping; and the source materials of the local roads are made of clayey materials which are susceptible to water. According to the current regulations, periodic maintenance has an intervention interval of every 4 years. However, as these roads use clayey materials that are susceptible to water, they deteriorate rapidly, one of the main factors being rainfall (climatological factors). The effectiveness of this type of intervention is less than 10 months, which generates costs to the state and deficiencies in service levels. El Vía Vecinal: Lircay - Jatumpata - Mitoccasa -Dv. Ccollpa - Pitinpata

- Perccapampa, Province of Angaraes, Department of Huancavelica, is a road from lowù volume road of traffic with a IMDA from 54vehicles/day , which has a length of 16+670 km. This road has a 15 cm thick pavement, with a platform width of 4.00 m with berms at the ends of 0.50 m, as well as ditches next to the slope of the land of a width of 0.75 m by 0.30 m in height.

In 2019, periodic maintenance was carried out on this road, which soon eroded rapidly due to heavy rainfall, which caused the aggregates to break up, potholes, rutting and rutting. In February 2020, routine maintenance began, where it was verified that after a few months the structure of the pavement had deteriorated: At present, the service levels of this neighbourhood road are in very poor condition, generating discomfort among the transporters and inhabitants of the area, which is why it is proposed to use a basic solution (polyacrylamide polymer PAM) as recommended in the technical document on basic solutions for unpaved roads to increase the properties of the soil due to the rapid deterioration of the road surface due to the effects of traffic and climate; and also the criterion of reducing the use of materials from other places.

1.2. Problem formulation and systematisation

1.2.1. Problem general

How does the application of polyacrylamide polymer (PAM) influence soil stabilisation on local roads in the province of Angaraes - Huancavelica?

1.2.2. Specific problems

a. What is the behaviour of the Plasticity Index of soil stabilised with the application of polyacrylamide polymer (PAM)?
b. What are the bearing capacity (CBR) values obtained from soil stabilised with the application of polyacrylamide polymer (PAM)?
c. How would the application of polyacrylamide polymer (PAM) improve service levels?

1.3. Justification

1.3.1. Practice or social

The research will positively benefit the direct and indirect beneficiaries of the study area by improving the level of comfort for travellers, reducing travel time, reducing fares, and/or many other positive effects.
It will improve socially, because it has an impact on agricultural production, to take from the production centres to the market, making their income, family and social status profitable, in the present and in the

future.

1.3.2. Scientific or theoretical

The justification is that there are technical and scientific studies of new soil stabilisation technologies, which are only applicable to roads with high traffic volumes such as departmental and national roads. The importance of their application on low traffic volume roads is a research challenge. Demonstrating that they are more economically profitable over time compared to traditional methods, and with savings in operation and maintenance costs and the useful life of the project.

1.3.3. Methodological

Given the positive results of the research and with the application of the laboratory procedures used to demonstrate validity, they will serve for future research involving the application of Polyacrylamide polymers (PAM).

1.4. Delimitation

1.4.1. Space

In the present investigation, the spatial development of the neighbourhood road: Lircay - Jatumpata - Mitoccasa -Dv. Ccollpa - Pitinpata - Perccapampa.

- Region : Huancavelica

- Province : Angaraes

- District : Lircay

- Locality Lircay - Jatumpata - Mitoccasa - Dv.

Ccollpa - Pitinpata - Perccapampa

1.4.2. Temporary

The present research was developed in the year 2021, as well as taking precedent from nationally executed projects and research previously developed from 2015 to 2021.

1.4.3. Economic

The costs involved in carrying out this research, including field and laboratory work, will be borne entirely by the researcher.

1.5. Limitations

1.6. Objectives

1.6.1. Objective general

To determine the influence of the application of the polymer polyacrylamide (PAM) on soil stabilisation on local roads in the province of Angaraes - Huancavelica.

1.6.2. Specific objectives

a) To analyse the behaviour of the plasticity index of soil stabilised with the application of polyacrylamide polymer.
b) Evaluate the bearing capacity (CBR) values obtained from soil stabilised with the application of polyacrylamide polymer (PAM).
c) Explain how the application of polyacrylamide polymer (PAM) would improve service levels.

2.1. Background

2.1.1. Background National

(Villanueva Flores, 2017) In her thesis for her Master's degree in Road Infrastructure **entitled** "Proposal for stabilisation of low traffic volume roads in the highlands, above 2000 m.a.s.l., using anionic polyacrylamide, organosilane and a sulphonate" sets as **General Objective: To** propose options to improve the behaviour of soils based on experimental studies of low traffic volume roads. Applying the **methodology** based on an experimental design, due to the fact that different doses of each of the three soil stabilisers are incorporated into the soil stabiliser. **Results** were obtained where the soils combined with anionic polyacrylamide have a better behaviour as the clayey gravel with sand with a PI of 11, likewise showing an increase in bearing capacity. Finally **it concludes** that the CBR of the natural quarry material of 27.4% was increased to 86.3% with the polyacrylamide polymer. (Nesterenko Cortes, 2018) In his thesis for his Master's degree in Civil Engineering with mention in Road **entitled** "Performance of polymer-stabilised soils in Peru", sets as **General Objective: To** identify the results of soils tested in their natural state in relation to soils stabilised with PAM polymer under the same conditions. Applying the **methodology** based on the interpretation of the results of laboratory tests. Obtaining **results** where it was observed that in all the tested soils the percentage of CBR is above 45%. Finally, **it concludes** that the polyacrylamide polymer PAM is an alternative solution for soils with low bearing capacity, increasing its CBR by more than 20% on average and reducing the Optimum Moisture Content, generating a saving of water in the execution of the stabilisation. (CHAVEZ Pajuelo, 2018) In his undergraduate thesis **entitled** "comparative study using the additive PROES and CONSOLID for soil stabilisation in local roads", sets as **General Objective:** To evaluate the influence of the application of PROES and CONSOLID additives in the stabilisation of soils on local roads. Applying the **methodology** based on a quasi-experimental design. Obtaining **results** that when applying the PROES additive, its bearing capacity increased to 45.7% as opposed to the CONSOLID additive, which obtained a CBR of 36.2% at 95% of the MDS. Finally, **it concludes** that the PROES and CONSOLID additives considerably improved the mechanical properties of the soil obtaining a CBR of 45.7% and 36.2% compared to the natural soil which obtained 3.8% at 95% of the MDS. In addition, there was a reduction of the PI up to 50%. (HUAMANI Gamarra, et al., 2018) in their undergraduate theses **Entitled** "Application of Z-stabiliser with polymer in the increase in the value of the CBR of the material used as a pavement on the departmental road ap-103, Ullpuhuaycco - Karkatera bridge section (L= 14.050 kms) Abancay-Apurímac 2018" sets as **General Objective: To** determine whether there is an increase in the value of the CBR of the material used as a pavement with the application of Z stabiliser with polymer. Applying the **methodology** that responds to the experimental positivist paradigm. Obtaining **results** where the value of the bearing capacity of the unaltered sample was 15.44% and when the Z stabiliser with polymer was added, it increased to 18.57% at 100% of the Maximum Dry Density. Finally, **it concludes** that the application of the Z-stabiliser with polymer increases positively the penetration stress curve, which shows that there is an increase in the CBR value. (Baldeon Sauñe, 2019) In his undergraduate thesis **entitled** "Analysis of the use of silica sand in the stabilisation of the subgrade" sets as **General Objective: To** analyse the results of the use of silica sand in the stabilisation of the subgrade. Applying the **methodology** based on a quasi-experimental design in which the variables cannot be controlled or manipulated. Positive **results** were obtained where a combination of C-50% silica sand obtained a CBR of 15.50% compared to the subgrade material which obtained 2.8% CBR which is not suitable. Finally, **it concludes** that combining the subgrade material with silica sand improves the properties of the soil by increasing its bearing capacity, being an alternative solution for inadequate subgrades.

2.1.2. Background International

(Aguilar Castañeda, et al., 2015) In his undergraduate thesis **entitled** "Review of the state of the art of the use of polymers in soil stability", sets as **General Objective:** To carry out a search for techniques for soil improvement through the application of polymer at national and international level. Applying the **methodology** based on the collection of information on soil stabilisation issues. Obtaining **results** where the improved soil tends to increase the compressive strength and tensile strength. Finally, **it concludes that** the use of polymers as a stabilising product increases the physical and chemical properties of the treated soil, such as resistance, CBR, durability and others. (ALTAMIRANO Navarro, et al., 2015) In his monographic work to obtain the title of Civil Engineer **entitled "**Stabilisation of cohesive soils by means of lime in the roads of the community of San Isidro del Pegón, municipality of Potosí-Rivas", sets as **General Objective: To** stabilise with hydrated lime the cohesive soils of the roads of the Community of San Isidro de Pegón, Potosí. Applying the **methodology of** compilation and evaluation of the critical points of the roads. Obtaining **results** where due to the combination of hydrated lime and clayey soil this reacted exothermically producing a significant increase in the bearing capacity. Finally, **it concludes** that with a combination of 9% lime with the cohesive material, greater results are obtained in the expansion or swelling, achieving a reduction of 61%.(RAMOS vasquez, et al., 2019) In his undergraduate thesis **entitled** "Soil stabilisation by means of alternative additives", sets as **General Objective: To** analyse the properties of the subgrade, applying coal ash and lime as alternative additives. Applying the **methodology** based on the compilation of information on the treated soil. Obtaining **results** in which lime with a 10% proportion is the one that supports the greatest maximum effort and in relation to the price - quality is the best option while that for coal ash with a 40% proportion he obtained better results. Finally, **it concludes** that combining the proposed additives with the subgrade material improves the mechanical behaviour and is an option with respect to the strength-to-cost ratio.

2.2. Framework conceptual

2.2.1.Peru's road system

In Peru, roads are made up of a road system which are the National Road Network, the Departmental Road Network and finally the Neighbourhood Road Network, which perform the following functions:

2.2.1.1. Road Diagram

The road diagram shows geographic references to show the location of a road, indicating the route code, length and type of road surface; these are located especially in the most important towns that they link.

2.2.1.1.1. Neighbourhood Roads

It is a road that belongs to the local road system and is the responsibility of the Provincial Road Institute. They are used to connect various population centres. These roads have a low volume of traffic, and most of them are built at the level of the road surface.

2.2.1.1.2. Type of work to be executed

a) **Routine maintenance:** These are activities which are carried out on a permanent basis in order to preserve the service levels of the road. These activities are related to road cleaning, pothole patching, profiling and other works, which are carried out on a regular basis. (Ministry of Transport and Communications, 2016 p. 6).

b) **Periodic maintenance:** These are activities scheduled at certain intervals that are carried out to maintain service levels. These activities may involve the intervention of machinery or people for profiling, levelling, replacement of material, and works of art. (Ministry of Transport and Communications, 2016 p. 7).

c) **Rehabilitation:** These are activities whose execution is necessary to restore the road to its original characteristics, taking into account a new period of service (Ministry of Transport and Communications, 2016 p. 5).

d) **Upgrading:** These are activities in which their execution raises the standard of the road, involving substantial modification of the geometry and the transformation of a dirt road to a sealed road (Ministry of Transport and Communications, 2016 p. 5).

2.2.2. The Soil

Soil is formed from the physical weathering of rocks. This process, known as weathering, promotes the transport of weathered material which is then deposited to form alterite, from which the soil is strengthened by various processes. (BAÑON Blázquez, et al., 2000 p. 2). Soil, in Civil Engineering, is a discrete sedimentary layer of solid particles, resulting from the transformation of rocks, or soil that is transported by morphogenic agents supported by gravity as a directional force. (DUQUE, et al., 2002). Soil is the most abundant construction material in the world in civil engineering activities. When the engineer uses soil as a material for construction, he must choose a suitable soil to provide support for structures such as buildings, roads, bridges, etc. (DUQUE, et al., 2002 p. 11).

2.2.2.1. Soil properties

Soil has the fundamental properties that must be taken into account:

a) **Granulometry:**

It represents the particle size distribution of the aggregates by sieving according to the technical specifications (MTC Test E 107).

The analysis of soil granulometry aims to determine the proportions of the different constituent elements, which are classified according to their size.

A good granular arrangement ensures that the soil performs well under loads. It should have an adequate proportion of gravel to support the loads and a percentage of plastic fines to bind the soil materials together. (Ministry of Transport and Communications, 2014 p. 30).

According to the size of the soil particles the following terms can be determined:

b) **Plasticity:**

It is the property that soils have up to a certain moisture limit without disintegrating, this property depends exceptionally on fine materials (Ministry of Transport and Communications, 2014 p. 31).

Plasticity is obtained from the difference between the liquid limit (LL) and plastic limit (LP), also called atterberg limits, which allows us to classify the soil. (Ministry of Transport and Communications, 2014 p. 31).

c) **Group Index**

These are based in part on the consistency limits, this index is standardised by the AASHTO Standard commonly used to classify a soil. The group index is obtained by the following formula:Source: Manual de carreteras, MTC

The group index is in an interval of 0 or more, which will be positive. If the group index becomes negative and if the group index is greater than 9, it will not be usable for roads. (Ministry of Transport and Communications, 2014 p. 32).

d) Natural Moisture

An essential characteristic of soil is its natural moisture content, which is associated with the density, especially of fines (Ministry of Transport and Communications, 2014 p. 33).

f) Carrying Capacity (CBR)

It is the load bearing capacity as a function of soil type, compaction and moisture content. It evaluates subgrades, sub base and base to be used for a pavement or for structural fills.

2.2.2.2. Soil Classification

An adequate classification will give an idea of the behaviour of the soil to be used, based on its properties that are easy to determine. By knowing the granulometry and atterberg limits of a soil, the mechanical behaviour can be predicted. A correlation of the two most widespread classification systems, AASHTO and ASTM (SUCS), will be presented below (Ministry of Transport and Communications, 2014 p. 33).

2.2.2.3. Affirmed

Aggregate is the combination of 3 types of aggregate which are stone, sand and fines. It requires adequate percentages for a good wearing course.

Types of pavement:

a) **Type 1 pavement:** made of granular material, with a PI of 12 when justified. This type of pavement is used on low traffic volume roads of class T0 and T1 with an IMDA of up to 50 vehicles per day (Ministry of Transport and Communications, 2008).

b) **Type 2 pavement:** made of a granular material, with a PI of 12 provided that it is justified. This type of pavement is used on low traffic volume class T2 roads with an IMDA of 51 to 100 vehicles per day (Ministry of Transport and Communications, 2008).

c) **Affirmed type 3:** consisting of granular material, with a PI of 12 w h e n   justified. This type of pavement is used on low traffic volume roads of class T3 with an IMDA of 101 to 200 vehicles per day (Ministerio de Transportes y Comunicaciones, 2008).

These pavements shall comply with the following conditions:

Liquid Limit: 35% Max

CBR: 40% Min to 100% MDS

2.2.2.4. Soil stabilisation

It is based on the incorporation of a chemical product, which is mixed intimately and homogeneously with the soil to be treated. Soil stabilisation is defined as the improvement of the physical properties of a soil by means of mechanical processes and the incorporation of natural and synthetic chemical products. Such stabilisations are generally carried out on soils with inappropriate or poor subgrade, in this case they are called soil cement stabilisation, soil lime, soil asphalt and various other products. (NESTERENKO Cortes, 2018).

It is also applied on a sub-base, base or granular material, which even if it meets the conditions of having a certain CBR value, will be stabilised for The use of a higher quality material with a lesser thickness of granular layer. In general, the application of this criterion is for roads where there is considerable heavy traffic or even in sectors with less traffic, but whose conditions merit its execution (Ministry of Economy and Finance,

2015). Mechanical stabilisation is carried out through the application of additives that act physically or chemically on the properties of the soil. Among the most commonly used are lime and cement, but sodium chloride (salt), magnesium chloride, liquid asphalts, slag and chemical products are also used. The application of the latter will be in accordance with the MTC 1109- 2004 Technical Standard for Chemical Stabilisers (Ministry of Transport and Communications, 2008 p. 155).

2.2.2.4.1. Soil stabilisation of unpaved roads

The aim of soil stabilisation is to improve the physical and mechanical properties of a soil, so that materials of insufficient stability can be used as subgrade and granular layers (Ministry of Transport and Communications, 2008 p. 154).

2.2.2.4.2. Most commonly used stabilisation techniques

a) Granulometric stabilisation: consists of combining different soils to obtain a material with better admissible characteristics to be used as a subgrade or as a pavement (Ministry of Transport and Communications, 2008 p. 157).

b) Lime stabilisation

Soil-lime is obtained by the intimate combination of soil, lime and water. The lime used is mainly composed of calcium oxide (quicklime), obtained by calcination of limestone materials, or calcium hydroxide (slaked lime or hydrated lime). (Ministry of Transport and Communications, 2008 p. 158).

c) Cement stabilisation

This type of stabilisation is one of the most widely used stabilisation methods currently used, y is being implemented It has been used since Amies introduced it in 1917 to avoid the phenomenon of pumping of fines characteristic of rigid soils. By combining this type of particles, it was achieved that the water would not dissolve them, thus preventing the concrete slabs from slipping and subsequently breaking. (BAÑON Blázquez, et al., 2000 p. 22) The material called soil-cement is obtained by the intimate combination of a properly disaggregated soil with cement, water and other possible additions, followed by adequate compaction and curing. (Ministry of Transport and Communications, 2008 p. 160).

2.2.2.4.3. Basic solutions on unpaved roads

The basic solutions are technical, economic and environmental alternatives, which consist of the application of soil stabilisers to maintain their levels of service and prolong the maintenance intervals that allow the proper transit of vehicles. These basic solutions are applicable to unpaved roads, at the maintenance, rehabilitation, upgrading and construction levels.

2.2.3. Polyacrylamide polymer (PAM).

Polyacrylamide (PAM) is a synthetic organic polymer made from acrylamide monomers. Depending on the number of different monomers that make up the polymer, we have: homopolymers, copolymers, terpolymers and others.In the 1990's, Polyacrylamide (PAM) was introduced to the US market for application in soil erosion control and by the end of that decade it was used on 400,000 ha of irrigated land which was treated. Due to its low toxicity PAM has been used as a soil stabiliser and soil erosion minimiser making this polymer versatile and effective (Application of polyacrylamide as an alternative for the treatment of hydrocarbon contaminated soils, 2013).Commonly used stabilisers such as lime and cement require long curing times and copious amounts of additives at a significant price, so non-traditional stabilisers such as polymers, in this case the polymer Polyacrylamide called PAM, have gained more attention, as it is potentially more efficient in the field because the polymer has greater workability with stabilised soils during the construction process and sustainable, i.e, preserve the service levels of stabilised roads over time compared to conventional ones (NESTERENKO Cortes, 2018).

2.3. Definition of terms

• **Soil stabilisation:** Procedures to improve the properties of a soil through the application of chemical or natural products (Ministry of Transport and Communications, 2013).

• **Levels of Service:** These are indicators that quantify and qualify the condition of a road, which are used as admissible limits of a road. These indicators vary according to technical and economic factors in order to satisfy user needs such as safety, comfort and others (Ministry of Transport and Communications, 2013).

• **Polyacrylamide (PAM):** Synthetic organic polymer formed from acrylamide monomers (Application of polyacrylamides in landscaping and gardening, 2005).

• **Polymer: A** macromolecule particle, which is repeated throughout the molecule (LOPEZ Carrasquero, 2004).

2.4. Hypothesis

2.4.1. General Hypothesis

The application of polyacrylamide polymer (PAM) in the stabilisation of the soil of the local roads in the province of Angaraes - Huancavelica will have a positive influence on the physical-mechanical properties.

2.4.2. Specific Hypotheses

a) The plasticity index of the stabilised soil is reduced by the application of the polyacrylamide polymer (PAM).

b) The bearing capacity (CBR) values of stabilised soil will increase significantly with the application of polyacrylamide polymer (PAM).

c) Long-term service levels would be preserved with the application of polyacrylamide polymer (PAM).

2.5. Variables

2.5.1. Conceptual definition of the variable Independent variable (X):

- Polyacrylamide polymer (PAM): Synthetic organic polymer formed from acrylamide monomers.

Dependent Variable (Y):

- Soil Stabilisation: Improvement of the physical properties of a soil through mechanical processes and the incorporation of natural or synthetic chemicals.

2.5.2. Operational definition of the variable Independent variable (X):
• Polyacrylamide polymer (PAM)

One of the most remarkable characteristics of polyacrylamide polymer when combined with soil is that it increases the bearing capacity, reduces permeability and reduces erosion.

Dependent Variable (Y):

• Soil Stabilisation

Improvement of the physical-mechanical conditions determined in the laboratory, which are determined from the granulometry, bearing capacity (CBR), resilience modulus, modified Proctor, atterberg limits and permeability coefficient.

2.5.3. Operationalisation of variables

Table 1 Operationalisation of variables

VARIABLE	CONCEPTUAL DEFINITION	OPERATIONAL DEFINITION	DIMENSIONS	INDICATORS	INSTRUMENT	SCALE
X: application of the Polyacrylamide {PAM}	Synthetic organic polymers formed from acrylamide monomers.	One of the most remarkable characteristics of the polyacrylamide polymer when combined with soil is that it increases bearing capacity, reduces permeability and reduces erosion.	D1: stabiliser	I1: Stabiliser properties	Laboratory sheets	Reason
Y: Soil stabilisation	One of the The most remarkable characteristics of the polyacrylamide polymer when combined with soil are that it increases the capacity for support, reduces permeability and reduces erosion.	The process of the present research is to carry out the laboratory tests which are: granulometry, CBR, Atterbeg limits, Modulus of resilience, Modified Proctor, Permeability	D1: Physical Properties D2: Mechanical Properties	I1: Granulometry test I2: Consistency I3: Soil classification I4: Permeability I1: Carrying Capacity {CBR}. I2: Compactness	Information gathering sheet	Reason Reason

Source: Own

CHAPTER III
METHODOLOGY

3.1. Research method

The scientific method is the way in which research is carried out in order to manifest the forms of existence of the objectives, to systematise and deepen the knowledge obtained, to subsequently explain and demonstrate in the experiment and with the techniques of its application. (RUIZ, 2007 p. 6). This research is scientific, because it was carried out through systematised steps to acquire reliable knowledge through hypotheses and observation, by testing with experimentation.

3.2. Type of research

Applied research seeks the application or utilisation of knowledge to then implement and systematise research-based practice (MURILLO Hernández, 2008). The present thesis is an applicative type of research, because it looks for mechanisms and strategies that allow to establish a concrete objective, which is the stabilisation of the soil through the application of the polymer polyacrylamide.

3.3. Research level

Descriptive research is that which describes the data and characteristics of a population; however, explanatory research is in charge of looking for the reason for the facts through the cause-effect relationship. (MARROQUIN Peña, 2012). The research level of this research is descriptive - explanatory, as it establishes a description of the problem; and explanatory, which determines the causes and consequences of this problem, in order to draw conclusions.

3.4. Research design

The research design is quasi-experimental, as it is based on analysing the effect of the independent variable (Polyacrylamide polymer) on the dependent variable (soil stabilisation).

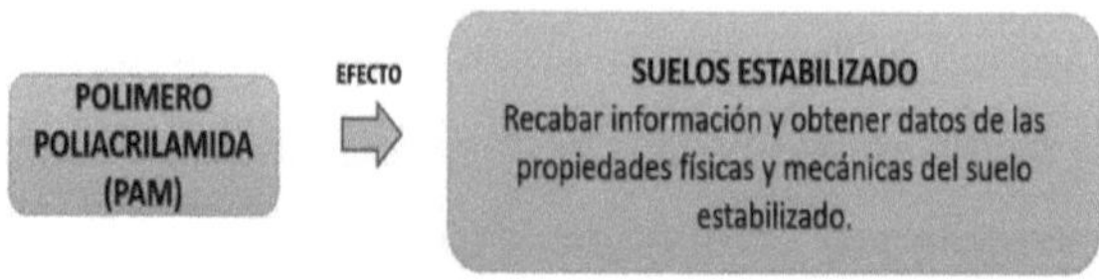

Figure 2 Research design Source: Own

3.5. Population and Sample

3.5.1. Population

The population is a group of individuals, objects or measures that have some similar characteristics visible in a given place and at a given time. (HERNANDEZ Hermosillo, 2013). According to the above, the population will be the local roads of the province of Angaraes, Huancavelica Region.

3.5.2. Sample

The sample is a characteristic subset of the population (HERNANDEZ Hermosillo, 2013). The sample consists of the Lircay - Jatumpata - Mitoccasa -Dv. Ccollpa - Pitinpata - Perccapampa local road, which has a length of 16 + 670 km.

3.6. Data collection techniques and instruments

3.6.1. Data collection techniques

Data collection is any technique used by the researcher to approach phenomena and obtain information from them (SABINO, 1992 p. 108). Sampling was carried out in the main quarry located on the right hand side of the road which will later be stored in waterproof bags. The instruments to be used will be soil mechanics laboratory equipment, photographic equipment, field notebook and computer equipment for data processing.

3.6.2. Validity

Validity represents the effectiveness with which an instrument measures what it is intended to assess (CHAVEZ, 2001). The instruments will be assessed by professionals who will be civil engineers.

3.7. Information processing

Once the sampling has been carried out, laboratory tests such as granulometry, determination of the atterberg limits, AASTHO and SUCS classification, Modified Proctor test and CBR test will be carried out according to the current regulations of the Ministry of Transport and Communications.

3.8. Techniques and data analysis

The data itself is of limited importance, it is necessary to "make it talk", which is the essence of data analysis and interpretation (ENCINAS Ramirez, 1993). The Research data will be processed and analysed as Frequency Distribution and Graphical Representations (Histograms; frequency polygons; bar charts; pie charts; cross tabulation). Normal distribution (normal test of percentages) of the data obtained from the granulometry, determination of the atterberg limits, AASTHO and SUCS classification, Modified Proctor test and CBR test (ENCINAS Ramirez, 1993).

4.1. Service level analysis

The levels of service recommended by the MTC which should be taken into account for an unpaved road, such as the road under study, are as follows:

- **Potholes:** The pavement should be free of potholes, these should be filled and compacted with borrow material.

- **Dust control:** The surface of the wearing course must have permanent dust control.

- **Replacement of pavement:** The minimum allowable thickness should be 15 cm.

- **Surface profiling:** Intervention interval is once a year.

For this reason, an assessment of the damage to the road surface was carried out, if, according to the MTC, it is complying with the levels of service that a paved road should provide.

4.1.1. Damage to the running surface

A verification of the Lircay - Jatumpata - Mitoccasa - Dv. Ccollpa - Pitinpata - Perccapampa local road was carried out, which has a length of 16+670 km where it was possible to see the different damages as detailed in the table of **DAMAGE TO THE ROLLING SURFACE:**

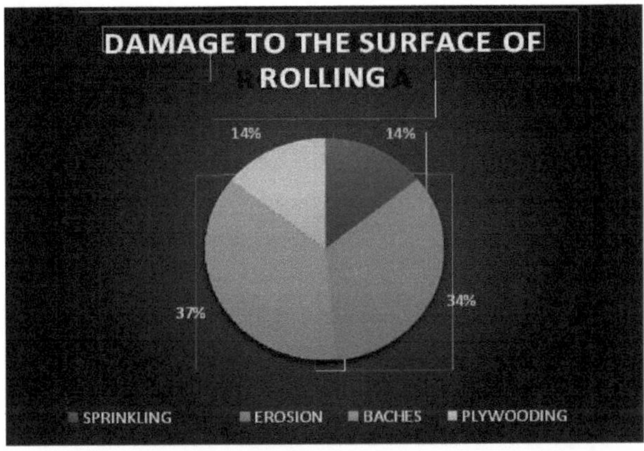

Figure 3 Percentage of damage on the local road

Source: own

A. Lircay - Jatumpata - Mitoccasa Progressive 0+000 KM to 12+000 KM

During the tour it was possible to verify that the structure of the pavement has suffered damage due to erosion, potholes and rutting; in this respect it can be stated that among the interval of the progressives, **damage 3 (potholes) has the** greatest impact.

Image 1 Presence of potholes Prog. 2+000 km Source: Own

B. Dv. Pitinpata Progressive 0+000 KM to 1+750 KM

The platform of the road to the village of Pintinpata shows a strong detachment of the aggregates caused by erosion, which according to the table of damage to the road surface would be **grade 2.**

Image 2 Detachment of aggregate Prog. 0+420 KM Source: Own

C. Dv. Ccollpa Progressive 0+000 KM to 1+600 KM

During a visit to the village of Ccollpampa, the presence of potholes was observed, which according to the table of damage to the road surface would be **grade 2.**

Picture 3 potholes Prog. 0+740 KM Source: Own

D. Dv. Perccapampa Progressive 0+000 KM to 2+000 KM During the

tour of this part of the section it was possible to verify that the structure of the affirmed has suffered damage due to rutting, potholes and rutting; in this respect it can be stated that **damage 3 (potholes)** was more prevalent.

Fig. 4 Aggregate liming and dispatch of aggregate Prog. 0+000 Source: Own

4.2. Analysis of granular material

4.2.1. Location

The entrance to the main quarry is located at Km 3+780 of the Lircay - Jatumpata road. It has the following UTM coordinates:

4.2.2. Accessibility

The quarry has a direct access of 40 m from the foot of the road to the collection centre.

Image 5 Access to the quarry Prog. 3+780 Km

4.2.3. Use

Based on the laboratory results and the current regulations of the unpaved roads manual, it is determined that this material will be used for periodic maintenance at the surfacing level.

4.2.4. Availability

The quarry belonging to the Comunidad Campesina San Juan de Dios is freely available.

4.2.5. Power of the quarry

The approximate area of exploitation of the granular materials as estimated is 6018.95 m2, with a stratum of 8.00 m on average. In relation to this, the net usable power was determined.

4.2.6. General characteristics of the material

The data obtained from the standard and special tests were based on the MTC's Manual of testing materials for roads and international standards, including statistical techniques for the analysis of the data, the classification of the material for the pavement was given using the SUCS and AASHTO methods, the tests of the atterberg limits (Liquid Limit, Plastic Liquid and Plasticity Index) were carried out, the modified Proctor test to establish the MDS and its OCH of the quarry material and finally the CBR test (Bearing Capacity) at 100% of the maximum dry density at penetration of the quarry material; Modified Proctor test to establish the MDS and its OCH of the quarry material and finally it was determined by means of the CBR test (Bearing Capacity) at 100% of the Maximum Dry Desity at a penetration of 0.1". **Table 3** General quarry material data

4.2.6.1. Consistency Limits

4.2.6.2. SUCS and AASHTO classification

SUCS (Unified Soil Classification System)

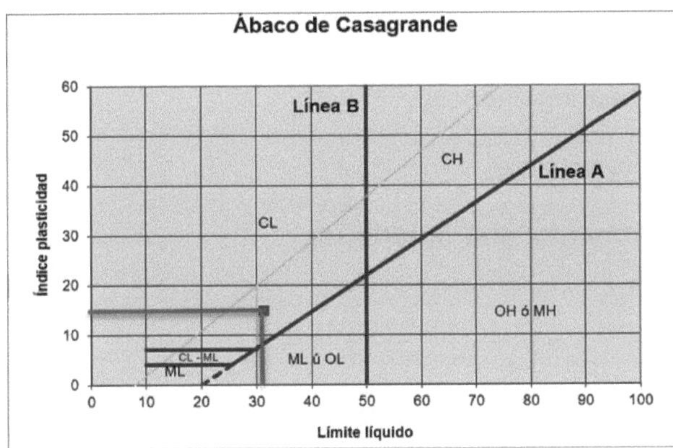

Figure 4 SUCS classification according to the Casagrande Abacus

Source: Obtained from Laboratory

AASHTO (American Association of State Highway and Transportation Officials)

$$IG=(F-35) \times (0.2+0.005 \times (LL-40)) +0.01 \times (F-15) \times (IP-10)$$

$$IG= (34.92-35) \times (0.2+0.005 \times (31.36-40)) +0.01 \times (34.92-15)^* (14.79-10)$$

$$IG= 1$$

Figure 5 Calculation of the Group Index of the quarry material

4.3. Analysis of the combination of quarry material and polyacrylamide polymer (PAM)

4.3.1. Polyacrylamide polymer (PAM)

4.3.1.1. Description

Polycom is an Australian chemical stabiliser which comes in the form of Acrylamide Powder Concentrate.

4.3.1.2. Technical characteristics

• Applicable on low quality soils

• Applicable for subgrade, sub-base and granular base as well as for road surface.

• Improves soil bearing capacity

• Increases soil resistance

4.3.1.3. Environmental characteristics

• Ecological

• Non-toxic

• Biodegradable

• Non-Flammable

• Non-hazardous product

Figure 6 PolyCom polyacrylamide polymer

Source: Own

4.3.2. General characteristics of the combination

The combination was carried out at 0.030 g x kg of material according to the supplier's technical specifications, which were tested for the atterberg limits (Liquid Limit, Plastic Liquid and Plasticity Index); modified Proctor test to establish the MDS (Maximum Dry Density) and its OCH (Optimum

Moisture Content) of the quarry material and finally it was determined by the CBR (Bearing Capacity) test at 100% of the Maximum Dry Density at a penetration of 0.1".

CHAPTER V

DISCUSSION OF OUTCOME

5.1. Influence of the application of Polyacrylamide Polymer (PAM).

In this research to determine the influence of the application of the polyacrylamide polymer (PAM) in the stabilisation of the soil, it was found that it improves the physical-mechanical properties of the stabilised material. This means that the material of the main quarry when combined with the polyacrylamide polymer (PAM) increases its bearing capacity, reduces the optimum moisture content, its plasticity index and also minimises damage to the surface; improving its service life and service levels of the road surface. In view of the above, the research hypothesis is accepted, which states that the application of polyacrylamide polymer (PAM) will influence the physical-mechanical properties in the stabilisation of the soil of the local roads in the province of Angaraes - Huancavelica. These results are corroborated by (Nesterenko Cortes, 2018) who in his research concludes that samples stabilised with PAM improve their mechanical properties compared to samples in their natural state. In this sense, by analysing these results, we confirm that by combining the polyacrylamide polymer (PAM) and the material of the main quarry, this influences acquiring greater strength and durability due to this will require less intervention at the level of periodic maintenance which would be an alternative solution to road problems.

Source: Obtained from laboratory

5.2. Behaviour of the Plasticity Index of stabilised soil

When analysing the behaviour of the plasticity index of the stabilised soil with the application of polyacrylamide polymer, it was verified that the stabilised sample presents a Plasticity Index of 11.94%, while the unstabilised sample presented a Plasticity Index of 14.79%; that is to say that the sample in its natural state with a PI of 14.79% is outside the TCM parameters, while the stabilised soil complies with the parameters, because it reduces the PI by 2.85%. In view of the above, the research hypothesis is accepted, which indicates that the plasticity index of the stabilised soil is reduced with the application of the polyacrylamide polymer (PAM). These results are corroborated by (Curitomay Najarro, 2018) who in his research concludes that the polymers as a chemical stabilizer produced a reduction in the plasticity index. By analysing the results we can determine that by applying the chemical stabiliser polyacrylamide (PAM) to a granular material it reduces the Plasticity Index.

Table 2 Plasticity index comparison

SOIL CLASSIFICATION		MATERIAL OF QUARRY	SOIL STABILISED	INDEX OF PLASTICITY
SUCS	AASHTO			
SC	A-2-6 {1}	X		14.79
SC	A-2-6 {1}		X	11.94

Source: Obtained from laboratory

Figure 7 Comparison of Consistency Limits

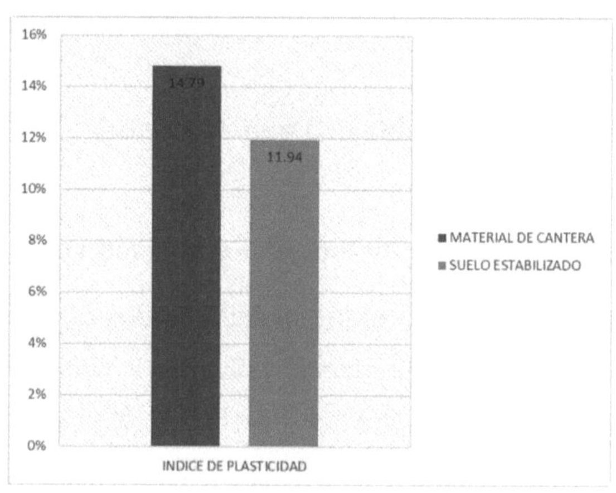

Source: Obtained from laboratory

5.3. Bearing Capacity Values (CBR)

When evaluating the bearing capacity (CBR) values obtained from the stabilised soil with the application of the polyacrylamide polymer (PAM) it was found that the CBR of the quarry material is 26.70 % at 100% MDS and the stabilised soil has a CBR of 43.10% at 100% MDS. This means that there is a 16.40% increase in bearing capacity (CBR) compared to the unstabilised quarry material. Therefore, the research hypothesis is accepted, which states that the bearing capacity (CBR) values of stabilised soil will increase significantly with the application of polyacrylamide polymer (PAM). These results are corroborated by

(Villanueva Flores, 2017) who in his research concludes that the CBR value increases significantly by 58.90% with the application of a polyacrylamide polymer in comparison with a natural material.

Table 3 Bearing Capacity Values

SOIL CLASSIFICATION		QUARRY MATERIAL	SOIL STABILISED	CBR 0.01% 100%MDS
SUCS	AASHTO			
SC	A-2-6 {1}	X		26.70
SC	A-2-6 {1}		X	43.10

Source: Laboratory Tests

Figure 8 CBR comparison

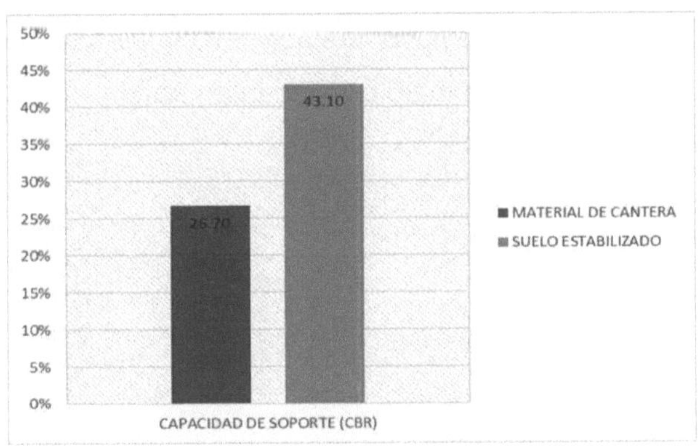

Source: Laboratory tests

5.4. service levels

By explaining how to improve service levels with the application of the polyacrylamide polymer (PAM), the service levels of the local road were evaluated, which show damage to the road surface such as rutting, erosion, potholes and liming, due to the fact that the material used for periodic maintenance is susceptible to water, The stabilisation of the quarry material with the polymer polyacrylamide (PAM) allows for better performance of the wearing courses, which will minimise the aforementioned damage at any time of the year, prolonging their useful life and reducing maintenance costs, meeting the parameters of the service levels and providing the user with comfort, safety and economy. The research hypothesis is accepted, where it is mentioned that the service levels would be maintained in the long term with the application of polyacrylamide polymer (PAM). This result is corroborated with the project executed on the Chaglla - Panao - Pachitea Province - Huanuco Region of 30 km in which the conservation of the service levels of the road is visualised 8 months after having stabilised the soil with the polyacrylamide polymer at the level of periodic maintenance. In this sense, it is confirmed that once the soil has been stabilised with the polyacrylamide polymer (PAM), its service levels will be maintained in good condition, requiring less maintenance intervention.

Image 7 Final state of the Chaglla - Panao section of the road Source: AustLatin

CONCLUSIONS

1. It is verified that applying the polymer polyacrylamide (PAM) to the quarry material influences its physical-mechanical properties, improving its useful life and reducing its intervention interval (periodic maintenance).

2. A reduction of the plasticity index of the soil stabilised with the polyacrylamide polymer (PAM) compared to the quarry material of 2.85% is verified. Therefore, the polymer polyacrylamide (PAM) has the ability to reduce the plasticity of a material, which most materials in their natural state do not meet the parameters of MTC.

3. It is verified that stabilised soil has a CBR of 43.10% at 100% MDS while the soil in its natural state has a CBR of 26.70% at 100% MDS. This means that there is an increase of 16.40% of the bearing capacity.

4. The application of polyacrylamide polymer (PAM) provides granular layers with improved performance, which will reduce damage such as rutting, erosion, potholes and rutting at any time of the year. This means that service levels will be met and the service life of the road will be extended, providing the user with comfort, safety and economy.

RECOMMENDATIONS

1. It is recommended that when determining the bearing capacity (CBR), the polyacrylamide polymer stabilised sample is cured for 28 days, to verify the best results according to the technical specifications of the product.

2. It is recommended that road improvement and periodic maintenance projects apply basic solutions to maintain service levels, to minimise operating costs and to involve better use of state resources.

3. It is recommended that an investigation of the condition of the running surface of a road stabilised with polyacrylamide polymer (PAM) at the level of the pavement be carried out.

REFERENCES

Aguilar Castañeda, Catherin Gisella and Borda Riveros, Yeraldin. 2015. State of the art review of the use of polymers in soil stability. Santos Tomas University, Bogota : 2015.

Aguilar, Catherin. State of the art review of the use of polymers in soil stability. Universidad Santo Tomas, Bogata : s.n.

ALTAMIRANO Navarro, Genaro Jose and DIAZ Sandino, Axell Exequiel. 2015. Stabilization of cohesive soils by means of lime in the roads of the community of San San Isidro del Pegón, municipality of Potosí-Rivas. Managua : Universidad Nacional Autonoma de Nicaragua, 2015.

Application of polyacrylamide as an alternative for the treatment of soils contaminated by hydrocarbons. **Osorio Bautista, Manuel and Adams Schroeder, Randy. 2013.** 2013, Kuxulkab, p. 83.

Application of polyacrylamides in landscaping and gardening. **Enseñat De Carlos, Luz and Cabot Moura, Orene. 2005.** 2005, Horticulture.

ARROYO Hilton, Nancy. 2012. Diseño y conservacion de pavimentos Rigidos. Universidad Nacional Autonoma de Mexico, Managua : 2012.

Baldeon Sauñe, Irvin P. 2019. Analysis of the use of silica sand in subgrade stability. Universidad Peruana Los Andes, Huancayo : 2019.

BAÑON Blázquez, Luis and BEVIA García, José F. 2000. Manual de Carreteras - Construccion y Mantenimiento. s.l. : Ortiz e Hijos, Contratistas de Obras, S.A., 2000.

BARRIOS Bolaños, Walter. 2007. Guía teórica y practica de los cursos de pavimentos y mantenimiento de carreteras. Guatemala : s.n., 2007.

CHAVEZ Pajuelo, Rafael Antonio. 2018. Comparative study using the additive Proes and Consolid for the stabilization of soils in neighborhood roads. Universidad Cesar Vallejo, Lima : 2018.

CHAVEZ, Alizo. 2001. Introducción a la Investigacion Educativa. 2001.

Curitomay Najarro, Carlos Jan. 2018. Stabilisation of clayey soils with copolymer type polymers, applied to medium traffic road works on the Pucaloma - Yanayacu - Socos District road. San Cristobal de Huamanga National University, Ayacucho : 2018.

DUQUE, Gonzalo and ESCOBAR, Carlos. 2002. Origin, formation and constitution of the soil. 2002.

ENCINAS Ramirez, Irma. 1993. Analisis de Datos. [On line] 1993. https://dspace.ups.edu.ec/bitstream/123456789/8180/1/UPS-QT06544.pdf.

HERNANDEZ Hermosillo, Silvia. 2013. Poblacion y muestra. Universidad Autonoma Del Estado De Hidalgo, Mexico : 2013.

HUAMANI Gamarra, Zayda and CONDORI Ñahuinlla, Visayda. 2018. Application of Z-stabilizer with polymer in the increase of the CBR value of the material used as pavement in the departmental road ap-103, Ullpuhuaycco - Ullpuhuaycco bridge section. Karkatera (l= 14.050 kms) Abancay-Apurímac 2018. Abancay : Universidad Tecnologica de los Andes, 2018.

LOPEZ Carrasquero, Francisco. 2004. Fundamentos de polimero. Merida : s.n., 2004.

Marroquin Peña, Roberto. 2012. Reliability and Validity of research instruments. Universidad Nacional De Educacion Enrique Guzman Y Valle, Lima : 2012.

MARROQUIN Peña, Roberto. 2012. Metologia de la investigación.

Universidad Nacional De Educacion Enrique Guzman Y Valle, Lima : 2012.

Ministry of Economy and Finance. 2015. Methodological guidelines for the development of pavement alternatives in the formulation and social evaluation of public road investment projects. Lima : s.n., 2015.

Ministry of Transport and Communications. 2013. Glossary of frequently used terms in road infrastructure projects. Lima : s.n., 2013.

2013. Manual de Carreteras Suelo - Suelo, Geologia, Geotecnia y Pavimentos. Lima : s.n., 2013.

2008. Manual for the design of unpaved roads with low traffic volume. Lima : s.n., 2008.

2016. Reglamento Nacional de Gestion de Infraestructura Vial. Lima : s.n., 2016.

2014. Suelo, Geologia y pavimento - Section Suelos y Pavimentos. Lima : s.n., 2014.

MURILLO Hernández, W. 2008. La investigación cientifica. 2008.

Nesterenko Cortes, Darko. 2018. Performance of polymer-stabilized soils in Peru. Universidad De Piura, Piura : 2018.

NESTERENKO Cortes, Darko. 2018. Performance of polymer stabilized soils in Peru. Universidad De Piura, Piura : 2018.

RAMOS Vasquez, Juan David and LOZANO Gomez, Juan Pablo. 2019. Soil stabilization using alternative additives. Bogota : Universidad Catolica de Colombia, 2019.

RUIZ, Ramon. 2007. El metodo cientifico y sus etapas. Mexico : s.n., 2007.

SABINO, Carlos. 1992. El proceso de investigación. Caracas : s.n.,

1992.

SALINAS, Pedro J. 2012. Methodology of scientific research. Merida :
s.n., 2012.

Villanueva Flores, Silvia Monica. 2017. Proposal for stabilisation of low
traffic volume roads in the highlands, above 2000 m.a.s.l., using anionic
polyacrylamide, organosilane and sulphonate. Ricardo Palma University,
Lima : 2017.

CONSISTENCY MATRIX

TITLE: "APPLICATION OF THE POLYMER POLYACRYLAMIDE (PAM) FOR THE IMPROVEMENT OF LOCAL ROADS IN THE PROVINCE OF ANGARAES - HUANCAVELICA 2021.

PROBLEM	OBJECTIVES	HYPOTHESIS	VARIABLE	METHODOLOGY
GENERAL PROBLEM: How does the application of polyacrylamide polymer (PAM) influence soil stabilisation on local roads in the province of Angaraes - Huancavelica? SPECIFIC PROBLEMS What is the behaviour of the Plasticity Index of soil stabilised with the application of polyacrylamide polymer (PAM)?	GENERAL OBJECTIVE: To determine the influence of the application of the polymer polyacrylamide (PAM) on the stabilisation of the soil of local roads in the province of Angaraes - Huancavelica. SPECIFIC OBJECTIVES To analyse the behaviour of the plasticity index of soil stabilised with the application of polyacrylamide polymer.	GENERAL HYPOTHESIS: It will influence the physical-mechanical properties with the application of the polymer polyacrylamide (PAM) in the stabilisation of the soil of the local roads in the province of Angaraes - Huancavelica. SPECIFIC HYPOTHESIS: The plasticity index of the stabilised soil is reduced by the application of the polyacrylamide polymer	INDEPENDENT VARIABLE: Polyacrylamide polymer (PAM) DEPENDENT VARIABLE: Soil Stabilisation	RESEARCH METHOD: The research method is scientific, because it was carried out in orderly steps to acquire reliable knowledge through hypotheses and observation, by testing with experimentation. TYPE OF RESEARCH: The present thesis is an applicative type of research, because it looks for mechanisms and strategies that allow to establish a concrete objective which is the stabilisation of the soil through the application of the polyacrylic polymer. RESEARCH LEVEL Descriptive and explanatory research. DESIGN: It is a quasi-experimental design POPULATION

		(PAM).	
What are the bearing capacity (CBR) values obtained from soil stabilised with the application of polyacrylamide polymer (PAM)?	To evaluate the bearing capacity (CBR) values obtained from soil stabilised with the application of polyacrylamide polymer (PAM).	The bearing capacity (CBR) values of stabilised soil will increase significantly with the application of polyacrylamide polymer (PAM).	All local roads in the province of Angaraes, Huancavelica Region. SAMPLE The sample consists of the Lircay - Jatumpata - Mitoccasa -Dv. Ccollpa - Pitinpata - Perccapampa local road, which has a length of 16 + 670 km.
How would the application of polyacrylamide polymer (PAM) improve service levels?	Explain how the application of polyacrylamide polymer (PAM) would improve service levels.	Long-term service levels would be preserved with the application of polyacrylamide polymer (PAM).	DATA COLLECTION TECHNIQUES AND INSTRUMENTS: The main techniques used were quarry sampling. The instruments to be used will be soil mechanics laboratory equipment.
			INFORMATION PROCESSING
			Once the sampling has been carried out, laboratory tests such as granulometry, determination of the atterberg limits, AASTHO and SUCS classification, Modified Proctor test and CBR test will be carried out.
			TECHNIQUE AND DATA ANALYSIS:
			The research data will be processed and analysed as Frequency Distribution and Graphical Representations.

Printed by Books on Demand GmbH, Norderstedt / Germany